ANIMAL ECO INFLUENCERS

BEAVERS
Wetland Architects

MEGAN BORGERT-SPANIOL

Consulting Editor, Diane Craig, M.A./Reading Specialist

Super Sandcastle

An Imprint of Abdo Publishing
abdobooks.com

ABDOBOOKS.COM

Printed in the United States of America, North Mankato, Minnesota
102019
012020

 THIS BOOK CONTAINS RECYCLED MATERIALS

Design: Kelly Doudna, Mighty Media, Inc.
Production: Mighty Media, Inc.
Editor: Liz Salzmann
Cover Photographs: iStockphoto, Shutterstock Images
Interior Photographs: Alamy Stock Photo, pp. 14, 18, 19; Getty Images/iStockphoto, pp. 4, 10, 11, 16, 17, 22, 23; Shutterstock Images, pp. 3, 4, 5, 6, 7, 8, 9, 10, 11, 12, 13, 14, 15, 17, 18, 19, 20, 21

Publisher's Cataloging-in-Publication Data
Names: Borgert-Spaniol, Megan, author.
Title: Beavers: wetland architects / by Megan Borgert-Spaniol
Other title: wetland architects
Description: Minneapolis, Minnesota : Abdo Publishing, 2020 | Series: Animal eco influencers
Identifiers: ISBN 9781532191855 (lib. bdg.) | ISBN 9781532178580 (ebook)
Subjects: LCSH: Beavers--Juvenile literature. | Aquatic mammals--Behavior--Juvenile literature. |
    Animal ecology--Juvenile literature. | Lake ecology--Juvenile literature. | Wildlife habitats--Juvenile
    literature.
Classification: DDC 599.37--dc23

CONTENTS

ECO INFLUENCERS

What is an eco influencer? It is an animal that can change its ecosystem. All members of an ecosystem affect one another.

Gray wolves hunt elk in the forest. This means elk eat fewer plants. So, there are more plants for other wildlife to eat.

Parrotfish eat **algae** that grows on coral **reefs**. This keeps the reefs healthy.

Beavers are eco **influencers**. They build structures that create **wetlands**. The wetlands support other wildlife. Beavers are wetland **architects**!

THINK!

Can you think of other animals that help shape their ecosystems?

WATER DWELLERS

Beavers live in **South America**, **North America**, **Europe**, and **Asia**. They live in fresh water. They move between the water and the land around it.

Beaver lodges in a pond

A beaver lodge in a marsh

BEAVER HOMES

Lakes

Marshes

Ponds

Rivers

Streams

WHERE BEAVERS LIVE

NORTH AMERICA

EUROPE

ASIA

AFRICA

SOUTH AMERICA

AUSTRALIA

BEAVER RANGE

N
W E
S

SUPER SWIMMERS

Beavers move easily in water. This is because their bodies are built for swimming!

Beavers have wide tails. Their tails act like paddles. They help beavers steer in the water.

Beavers' noses and ears close underwater. This keeps water out.

Beavers have **waterproof** fur. It keeps them warm and dry.

Beavers have clear **inner eyelids**. They act like goggles so the beaver can see underwater.

Beavers have large lungs. They allow the beaver to stay underwater for up to 15 minutes.

Beavers have **webbed** feet. Their feet are like swim fins. They push the beaver through the water.

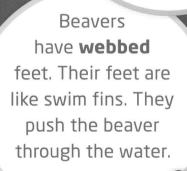

THINK!
How do the different parts of your body let you do daily activities?

BRANCH BITERS

Beavers cut down trees with their teeth. Beavers have strong teeth and jaws. This helps them chew wood. They use the wood to build dams and lodges.

NEW PLANTS, NEW ANIMALS

Beavers change their **ecosystem** when they cut down trees. This creates open spaces in forests. More sunlight reaches the forest floor. This helps new plants to grow. Other animals come to eat the new plants.

THINK!

What are some natural building materials outside your home or school?

BUILDING DAMS

Beavers use wood to build a dam across a stream. The dam blocks the flow of water. The water gets higher on one side of the dam. It creates a pond!

Beaver dams help shape a beaver's **ecosystem**.

Beavers fill the spaces between branches with mud.

Beaver dam

Beaver dams create ponds that support life.

The ponds are good for plants and animals that live in calm water.

Beaver dams help clean the water.

The dams trap **pollutants** that come from farms located near the stream.

Beaver dams reduce **erosion**.

A dam slows the stream's flow. Less dirt gets pulled from the banks. And more dirt in the water settles to the bottom.

BUILDING LODGES

Beavers build lodges in the ponds they create. Beaver families live in the lodges.

Like dams, lodges are made of logs, branches, and mud. The entrance to a lodge is underwater. Inside, there is a room above the water. This is where the beavers sleep and eat.

EATING CHAMBER

NESTING CHAMBER

UNDERWATER ENTRANCES

Beaver lodges also help shape the **ecosystem**. They create **habitats** for other animals.

In winter, beavers often share their lodges with other animals. These animals include frogs and **muskrats**.

Canada geese and other birds sometimes build nests on top of beaver lodges.

Muskrat

Canada goose

THINK!
What materials is your home made of? What creatures live in or around your home?

FOOD WEB

Beavers also affect their **ecosystems** through the **food web**. Beavers eat certain organisms. They are eaten by other organisms.

Beavers eat mainly poplars, birches, and willows. Beavers cut down these trees for food. This causes other types of trees to become more common.

Hawk

Coyote

Bear

Owl

River otter

Wolf

BEAVER PREDATORS

Beavers are prey to several animals. As prey, beavers support these animals in the **ecosystem**.

THINK!

What plants or animals do you eat? Where does your food come from?

WETLANDS AND MEADOWS

An area with beavers looks very different from one with no beavers. The beavers build dams. The dams create **wetlands**. Wetland animals join the **ecosystem**.

A wetland created by a beaver dam

A beaver pond turning into a beaver meadow

Beavers continue to **influence** their **ecosystem** long after they leave it.

AFTER BEAVERS LEAVE A HABITAT...

The beaver dam breaks.

The water flows out of the beaver pond.

A meadow grows where the bottom of the pond was.

This is known as a beaver meadow!

THINK!

Draw a picture of a forest stream. Now draw what the habitat would look like if beavers lived there.

PROTECTING BEAVERS

Humans are the greatest danger to beavers. In the 1800s, humans hunted beavers for their fur. Beavers nearly died out.

Today, hunting beavers for fur is less common. But many landowners see beavers as pests. Beavers can harm trees and cause flooding. Some people hunt or trap beavers to remove them from the land.

Wildlife **experts** know beavers are important. These people safely move beavers to areas that need **wetland architects**!

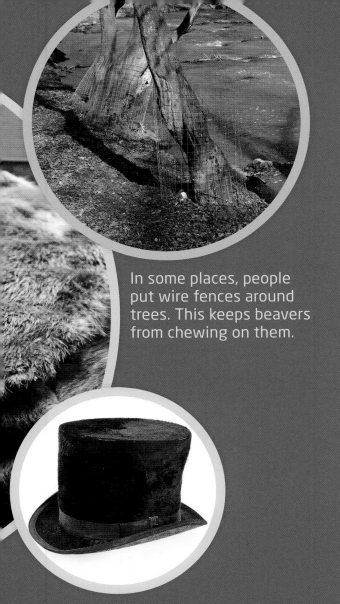

In some places, people put wire fences around trees. This keeps beavers from chewing on them.

Beaver fur was often used to make top hats.

ECO INFLUENCER FACT SHEET

Common name: **Beaver**

Class: **Mammal**

Life span in the wild: **Up to 24 years**

Population trend: **Stable**

Diet: **Herbivore**

Size in relation to humans:

FUN FACTS

A beaver family is called a colony. Members of a colony work together to build their lodge.

The world's largest beaver dam is in Canada. It is 2,790 feet (850 m) long. It can be seen from space!

A beaver can bite through a willow branch the size of a person's finger.

BEAVER QUIZ

1. What do beavers build to block a stream?
    A. dam
    B. lodge
    C. bridge

2. Beaver dams help clean the water.
    True or **false**?

3. What does a beaver pond turn into after beavers have left?

GLOSSARY

algae—plant or plantlike organisms that live in water.

architect—someone who designs and builds structures.

Asia—the largest of the continents. Russia, India, and China are in Asia.

ecosystem—a group of plants and animals that live together in nature and depend on each other to survive.

erosion—wearing away of the land often caused by water or wind.

Europe—the continent between Asia and the Atlantic Ocean. England, France, and Italy are some of the countries in Europe.

expert—a person very knowledgeable about a certain subject.

eyelid—the skin and muscle that can be closed over the eyeball.

food web—the feeding relationships between different organisms in a community.

habitat—the area or environment where a person or animal usually lives.

influence—to cause something to change.

inner—on the inside.

material—something that other things can be made of, such as wood, plastic, or metal.

muskrat—a North American rodent that lives in or near water and has webbed hind feet, a long scaly tail, and dark brown fur.

North America—the continent that includes Canada, the United States, and Mexico as well as other countries.

pollutant—a substance that makes air, water, or land dirty or not safe.

reef—a strip of coral, rock, or sand that is near the surface of the ocean.

South America—the continent surrounded by the Pacific Ocean, the Atlantic Ocean, and the Caribbean Sea. Colombia, Brazil, and Peru are in South America.

waterproof—made so that water can't get in.

webbed—having skin connecting the fingers or toes.

wetland—a low, wet area of land such as a swamp or a marsh.